Horse Colors

written by Mia Coulton
photographed by
Mia Coulton & Hannah Brown

This horse is black.

This horse is white.

5

This horse is gray.

This horse is brown.

9

This horse

is brown and white.

"Neigh."

To
Oliver,
Merry Christmas!
Love,

· ·

'Twas the night before Christmas,
when all through the house,
Not a creature was stirring,
not even a mouse;
Oliver's stocking was hung
by the chimney with care,
In hope that St. Nicholas
soon would be there.

Oliver was nestled all snug in his bed,
While visions of candy canes danced in his head.
And Mom in her kerchief, and Dad in his cap,
Had just settled down for a long winter's nap.

When out on the street
there arose such a clatter,
Oliver sprang from his bed—
what was the matter?
Up to the window
Oliver flew like a flash,
Tore open the curtains,
threw open the latch.

The moon on the blanket
of new-fallen snow,
Shone bright as midday
on the objects below—

When, what to Oliver's
wondering eyes should appear,
But a miniature sleigh
and eight tiny reindeer.

With a little old driver,
so lively and quick,
Oliver knew in a moment
it must be St. Nick.
More rapid than eagles
his reindeer they came,
And he whistled, and shouted,
and called them by name:

To
Oliver

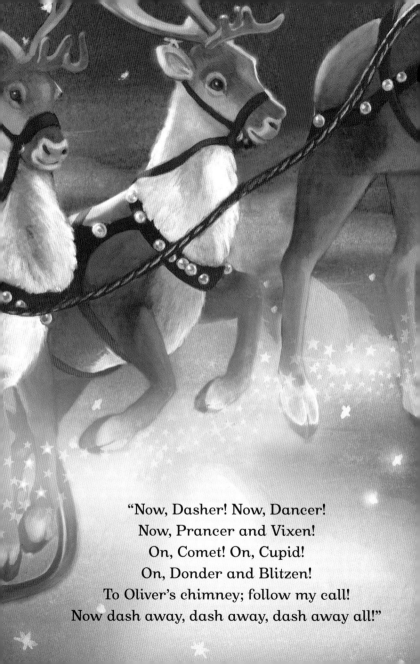

"Now, Dasher! Now, Dancer!
Now, Prancer and Vixen!
On, Comet! On, Cupid!
On, Donder and Blitzen!
To Oliver's chimney; follow my call!
Now dash away, dash away, dash away all!"

Then, in a twinkling,
Oliver heard on the roof
The prancing and pawing
of each little hoof.

Oliver pulled in his head
and was turning around,
When down the chimney
St. Nicholas came with a bound.

He was dressed all in fur,
from his head to his foot,
And his clothes were all tarnished
with ashes and soot.
A bundle of toys he had
flung on his back,
And he looked like a peddler
holding his pack.

To
Oliver

His eyes—how they twinkled!
His dimples—how merry!
His cheeks were like roses,
his nose like a cherry.
His droll little mouth
was drawn up like a bow,
And the beard on his chin was
as white as the snow.

Dear Santa,
I hope you enjoy
the sweet treats.
Love,
Oliver
P.S. The carrots
are for your
furry friends.

A big sack of toys
he held tight in his fist;
He glanced to see Oliver
on top of his list.
He had a broad face
and a little round belly
That shook when he laughed,
like a bowl full of jelly.

Oliver

NICE LIST

Oliver

Steven

Glenn

Marcus

Lorenzo

He was chubby and plump,
a right jolly old elf;
Oliver laughed when he saw him,
in spite of himself.

St. Nick winked an eye
and tilted his head
To let Oliver know
he had nothing to dread.

He spoke not a word,
but went straight to his work,
Filled Oliver's stocking,
then turned with a jerk.

And tapping his finger at the side of his nose,
And giving a nod, up the chimney he rose.

He sprang to his sleigh, to his team gave a whistle,
And away they all flew like the down of a thistle.
But St. Nicholas exclaimed, as he drove out of sight—

"Merry Christmas, Oliver, and to all a good night!"

Oliver, draw all your favorite Christmas presents from Santa.

Adapted from the poem by Clement C. Moore
Illustrated by Lisa Alderson
Designed by Jane Gollner

Copyright © Bidu Bidu Books Ltd 2021

Put Me In The Story is a
registered trademark of Sourcebooks, Inc.
All rights reserved.

Published by Put Me In The Story,
a publication of Sourcebooks, Inc.
P.O. Box 4410, Naperville, Illinois 60567-4410
(630) 536-1104
www.putmeinthestory.com

Date of Production: June 2021
Run Number: 5022128
Printed and bound in Italy (LG)
10 9 8 7 6 5 4 3

MIX
Paper from
responsible sources
FSC® C023419

Bestselling books starring your child!
www.putmeinthestory.com